Brunel's Bristol

Brunel's Bristol

R. A. Buchanan
and
M. Williams

REDCLIFFE
Bristol

First published in 1982 by
REDCLIFFE PRESS LTD
49 Park St, Bristol BS1 5NT

Reprinted 1992

ISBN 0 905459 45 8

Designed by Nicholas Bassett
Printed by The Longdunn Press Ltd, Bristol

Contents

Cover illustration: *The Launching of the Great Western 1837,*
an impression by the marine painter Arthur Wilde Parsons.

Acknowledgments

The text of this book is based on my essay 'Brunel in Bristol', in Patrick McGrath and John Cannon (eds): *Essays in Bristol and Gloucestershire History*, the centenary volume of the Bristol and Gloucestershire Archaeological Society, Bristol, 1976, but has been considerably modified and extended. As on the occasion of the publication of the original essay, I am glad to express my gratitude to Mr. N. Higham, Bristol University Librarian, and to Mr. G. Maby, Bristol University Archivist, for their continuing help during my research on the Brunel Collection in the Library of the University of Bristol. Of the papers in this collection, the Private Letter Books covering the years 1836 to 1859 are referred to hereafter as PLB. I would also like to thank the many audiences in Britain and Australia who have heard me talk about Brunel and have suggested many improvements which I have endeavoured to incorporate in this text.

R.A.B.

Introduction

I. K. Brunel was not a Bristolian, but the fact that most of his great engineering achievements were based in Bristol has meant that, with the rising tide of public interest in Victorian engineering works, he has come to be adopted as one of the favoured sons of Bristol. It has not always been so. Brunel's enterprises were striking for their daring innovations, and such novelties were frequently expensive for the investors in his projects, so that they met sharp criticism in his lifetime and thereafter. The Bristol chronicler John Latimer, writing at the end of the nineteenth century, was not untypical in dismissing Brunel as 'an inexperienced theorist, enamoured of novelty, prone to seek for difficulties rather than to evade them, and utterly indifferent as to the outlay which his recklessness entailed upon his employers'.[1] Such bitterness is a useful reminder of the problems facing the innovative engineer at any time, and especially in that unique period of some three decades in the middle of the nineteenth century when the advent of the railways transformed the profession of engineering from that of a fairly small group of experts, involved mainly in civil engineering, into a large-scale business of specialists active in many diverse fields of professional skill. For a few years, engineers of genius like Brunel were able to range widely over civil, mechanical, and marine engineering. His great contemporaries such as Robert Stephenson and Joseph Locke shared in this opportunity to practise as engineering polymaths. None of them, however, ranged so widely or with such visionary imagination as the younger Brunel.

Isambard Kingdom Brunel was born in Portsea on April 9th 1806, while his father, Marc Isambard Brunel, was engaged on the block-making machinery commissioned for the Admiralty at Portsmouth Dockyard. This commission had brought Marc Brunel from the United States where he had sought refuge, as a French Royalist, from the Revolutionary forces which had precipitated Europe into war and thus, incidentally, created a large demand for the rigging-blocks used in hundreds on the ships of the Royal Navy. The young French engineer had devised an ingenious method of mechanically mass-producing these blocks on what was, in effect, a production-line system, and he had been employed by the Admiralty to install it. On arriving in Britain in 1799, he had taken an early opportunity of renewing the acquaintance of an English lady, Miss Sophia Kingdom, whom he had encountered when both were fleeing from the Terror in France six years before. They were quickly married, and their only son was born in 1806 and named Isambard Kingdom — his father's second name and his mother's maiden name respectively.

When the French Wars ended in 1815, the demand for rigging-blocks collapsed and Marc Brunel turned his attention to other engineering ventures, none of which were financially successful, until he became involved in the scheme to construct a tunnel under the River Thames in London. Even though the family was desperately short of money, however, the young I. K. Brunel was sent to France to complete his education in 1819, and on returning he joined his father as Assistant Engineer on the Thames Tunnel project, soon becoming the Resident Engineer responsible for supervising the work on the site according to his father's designs. In January 1828, he narrowly escaped with his life when the Thames broke in and flooded the half-finished tunnel. He sustained internal injuries which, although not specified in any of the surviving accounts of the accident, were sufficiently serious to make his parents send the young man on a long convalescence, first to Brighton, which they soon came to regard as unsuitable for their son's recuperation, and then on to Clifton, the elegant suburb which was then developing on the edge of open downland to the north-west of Bristol.

There is a problem about the timing of Brunel's first visit to the Bristol region. His father kept a diary which was both more meticulous and more legible than any personal papers written by his son, and in the year following the disastrous 'accident at the Tunnel' which was recorded on Saturday January 12th 1828 there are frequent references to Isambard's state of health.[2] But the only occasion in the course of that year when he mentions Isambard leaving town was on July 12th, when the entry reads: 'Isambard has set off for Plymouth in the steam boat', and the first mention of Bristol is in the entry for January 23rd 1830: 'Isambard set off for Bristol'. Nevertheless, Brunel's biographers, with sources at their disposal which are not at present available, are clear about Isambard's visit to Bristol in 1828, even though they give no particulars about its date, its length, where it was spent, or whether or not he went with any introductions which would have put him in touch with the mercantile élite of Bristol.[3] The entries in Marc Brunel's diaries suggest that, in any event, the visit was not a protracted one.

Isambard was certainly back in Bristol the following year, because by then he had heard of the announcement by the Bristol Society of Merchant Venturers of the decision to build a bridge over the Avon Gorge which would thus link Clifton with Leigh Woods on the Somerset side of the river. A competition for designs was held in 1829, and on the closing day – November 19th – Brunel submitted four schemes of possible bridges within the terms of the competition, all beautifully drawn. By becoming involved in this competition, Brunel came into contact with a group of influential Bristolians, many of whom became his life-long friends, at a time when the city was poised on the threshold of a period of vigorous commercial and industrial activity. It was natural that, despite his tender years, Brunel's great talents should have been recognized and called upon by this group when harbour works, railway projects, and pioneering ship-building enterprises were being mooted in the city.

Although from 1829 onwards Brunel was constantly involved with engineering commitments in Bristol for at least twenty years, he never made his home in the city. He remained based in London, and when he married and set up a family home of his own it was over his office at 18 Duke Street, overlooking St. James's Park and convenient for Parliament and Whitehall. When he acquired a country estate he chose to go beyond the Bristol region, purchasing a property at Marychurch near Torquay in Devon close to his South Devon Railway, and even though he did not live long enough to settle there he devoted a lot of attention over many years to preparing the house and its gardens.

From 1835, when the Great Western Railway Company was established, until his death in 1859 at the age of 53, Brunel spent little time in Bristol. Most of his visits to attend Board meetings and to inspect his various works seem to have been hurried affairs of one or two nights. There are no details in the available records about where he stayed, but it could have been at the GWR office at Temple Meads Station, when this was completed in 1841, although it would seem more probable that he spent it in the special carriage which he had constructed for his personal use. In any event, it is clear from the frequent complaints of shareholders and other interested parties in Bristol who thought that he was neglecting their affairs, that he could not have spent long in the city during these years.

Yet despite his lack of any residential qualification, Brunel acquired a special affection for Bristol. He referred on one occasion in 1845 to 'our revered parent the City of Bristol'[4] and later in the same year protested to a Bristol colleague that: 'I have no wish in the matter whatever except the old and strong wish of being considered a Bristol man and one who can always be relied on as sticking to his friends thru' thick and thin'.[5] A decade later he wrote warmly of 'the spirited merchants of Bristol' who had led the world in steam navigation.[6]

There can be little doubt that Brunel's continuing loyalty to Bristol even at times when official relations had become strained and less than cordial was due to the close attachment which he formed with a group of leading Bristol industrialists and businessmen. These personal friendships were crucial to Brunel's links with Bristol. In them, he found men who appreciated his dynamism and who shared his visions. Through them, he was invited to undertake his many enterprises in Bristol. And when, at last, the group dispersed in the last decade of Brunel's life, so his connection with Bristol and its affairs also dwindled away.

The Bristol circle with whom Brunel became involved in the 1830s and 1840s included Tory merchants like Nicholas Roch, Liberal industrialists like T. R. Guppy, shipbuilders like William Patterson, and Captain Christopher Claxton, a retired naval officer employed as Quay Warden in the Bristol Docks. He also came to know men such as Osborne, Secretary to the Bristol Docks Company and a prominent legal adviser to the Society of Merchant Venturers and all manner of industrial enterprises. It appears to have been Roch who introduced Brunel to the Board of the Bristol Docks Company, and also to the promoters of the Great

Western Railway. Although Roch seems shortly afterwards to have retired to Pembrokeshire, Brunel was anxious in later years not to cause any offence to his old friend when the South Wales Railway was being built. We find him in 1844 promising to call on Roch 'to renew a much valued and *now* I may call it an old friendship',[7] and later in the year he instructed his assistant Brodie, surveying the line near Pembroke, to 'try and keep the line a little further from Mr Roches'.[8] His warning apparently came too late because a few days later he wrote apologetically to Roch: 'a singular chance or fatality has carried my levels almost thru' your house . . .'.[9] Brunel never forgot that Roch and the other Bristol men had given him his big opportunities as an engineer. These were 'the spirited merchants of Bristol' who chose him as their agent to accomplish the great designs which matured in the city in the 1830s.

These designs represented a belated but determined attempt to arrest the comparative decline of Bristol as a major centre of national enterprise. The reputation of Bristol had been going down since the early eighteenth century, at which time it had been second only to London amongst the seaports and cities of the realm. The vigorous growth of rivals, particularly Liverpool, had combined with a certain complacency amongst Bristolians to produce a situation of relative stagnation in the trade of the city. The increasing volume of national trade and the growth in the size of ships served to make the port of Bristol, consisting of riverside wharves in the heart of the city, some eight miles from the open sea up the tortuous and tidal River Avon, seriously inadequate. When at last these harbour facilities were improved by the Bristol Docks Company (BDC) between 1804 and 1809, the problems were only partially alleviated. This was because the improvements had proved to be much more expensive than had been envisaged by William Jessop, the engineer who had designed them, and also because the BDC imposed high harbour dues in order to recoup their losses. Thus traffic had no encouragement to switch from Liverpool and elsewhere, and a sense of frustration mounted in Bristol in the years following the Napoleonic Wars, as its more able citizens tried in vain to revive the drooping fortunes of the city.

This frustration became directed against the commercial oligarchy of the corporation and the unreformed parliamentary representation, culminating explosively in the Bristol Riots of 1831. The three days of chaos at the end of October 1831 have been described as: 'the worst outbreak of urban rioting since the Gordon Riots in London over fifty years earlier'.[10] At least twelve rioters died, in addition to those who were later hanged, and large parts of the city centre were laid waste. The Bristol Riots have posed a problem for historians because of the strange detachment from the turmoil shown by the well-to-do middle classes in Bristol, at least for the first two days, and because of the ambivalent attitude of the troops brought in to restore order. The conclusion of one recent study is that: 'There can be no doubt that it was the fate of the city Corporation, and not that of the Reform Bill, which was really at issue in Bristol in 1831'.[11]

The riots demonstrated the intensity of local feeling against the self-appointed oligarchy of wealthy merchants who kept a tight control over Bristol through the unreformed Corporation and the Society of Merchant Venturers. There was widespread relief when the power of this clique was modified if not entirely broken by the Municipal Corporations Act which became law at the end of 1835 and which established in Bristol an elected council with effective powers to organize a police force and to perform other vital functions.[12] These reforms had become an essential prerequisite of the sort of industrial expansion envisaged by Brunel and his friends.

Brunel was present in Bristol at the end of the riots, and gave evidence at the subsequent trial of the Mayor, Charles Pinney. But although it is intriguing to speculate about his part in the affray, this cannot be done with any certainty.[13] As he spent part of the critical three days with his friends Alderman Hillhouse and Nicholas Roch, it can be taken for granted that he did not sympathize with the mob violence. On the other hand, it seems likely that he would have shared in the popular animosity against the closed corporation. The next year he was out on the hustings in Lambeth supporting his brother-in-law Benjamin Hawes, who successfully contested the seat in the Radical interest for a place in the newly reformed House of Commons. Generally, however, Brunel was reserved about declaring any partisan political interest, and no evidence has so far come to light of his expressing any overt party point of view in Bristol affairs. In this he behaved very much in the tradition of British professional engineers and public servants. By shunning active political commitments he managed to retain the friendship of men with a variety of different political points of view.

The initial impact of the reform agitation, and especially the havoc caused by the Riots in 1831, was not conducive to commercial vigour in Bristol. Even the Clifton Bridge scheme was allowed to languish until confidence began to revive. In the autumn of 1832, Brunel was in a mood of deep despondency about his career, as none of the projects in which he was involved were making any progress. But when plans for a railway link between Bristol and London began to stir at the end of that year, they initiated a general revival of activity which quickened in the following years. Between 1833 and 1848 Bristol was the scene of a remarkable burst of political and commercial enterprise. The municipal government was reformed, giving the new industrial middle classes the same sort of enlarged scope in local affairs that the reform of parliament had given them at national level. At the same time, the national agitation for free trade, in which the same social classes were taking a leading part, was represented in Bristol by a movement to liberalize trade in the docks by reducing dues and, if necessary to achieve this, by acquiring the Docks Company for the Corporation, a transfer which was performed in 1848.

On the commercial and industrial front, entrepreneurial zeal was expressed in a remarkable crop of new enterprises, with railway works prominent amongst them.

The GWR was established with a capital of £2,500,000, increasing to over £8 million by 1844. It was followed by other railway companies, the most important of which were the Bristol & Exeter (established with a capital of £2 million in 1836), the Bristol & Gloucester (built in stages and opened in 1844), and the Bristol & South Wales Junction Railway (promoted in 1845 with a capital of £200,000, but not completed to New Passage ferry, after further injections of capital, until 1863). Brunel was involved to some degree in all these railway enterprises, and played a dominant role in most of them.

The railway boom stimulated other activities, many of which also called on Brunel's engineering skills. The prospectus for the Great Western Steamship Company appeared in January 1836 with a capital of £250,000, and its first vessel, the S.S. *Great Western*, was launched on July 19th 1837. The S.S. *Great Britain* followed in 1843, and Brunel was the designer of both ships. With the need for an oceanic terminal serving the land-locked port of Bristol, the Portbury Pier and Railway Company was floated in 1846 with a capital of £200,000 to carry out a scheme devised by Brunel. Even the impoverished Docks Company was inspired to put in hand much needed improvements, culminating in the new South Entrance Lock to Cumberland Basin, which became known as 'Brunel's Lock' after its designer. A brand new enterprise in Bristol was the Great Western Cotton Company, with a capital of £200,000 and a large factory in Barton Hill, the foundation stone of which was laid in April 1837. Conrad Finzel set up his sugar refinery, destined to become one of the largest in the country, on the Counterslip near the old Bristol Bridge in 1836. Fry's chocolate establishment, Christopher Thomas's soap enterprise, and Wills' tobacco company all underwent substantial growth in this period. E. S. Robinson set up his paper bag business in Baldwin Street in 1844, a year after William Butler became manager of the tar distillery at Crew's Hole which was later to take his name, and which owed its prosperity in part to Brunel's use of one of its products – creosote – for preserving railway timber.

These initiations of activities indicate the vigour associated with Bristol at the time when Brunel was most active in the affairs of the city. Unprecedented sums of capital were being raised for new enterprises, many of them promoted by friends of Brunel and all of them by people who knew him and who turned to him for engineering advice. The fact that Brunel's association with Bristol coincided with this extraordinary outburst of entrepreneurial activity may not have been accidental, as there is good reason to believe that Brunel's own enthusiasm had a galvanic effect on his friends and on enterprises in which he was involved. The association certainly had a most important formative effect on Brunel's own career. And it can be argued that the dismemberment of the association after 1848 had serious consequences for all the parties concerned.

By 1848, some of the relationships between Brunel and Bristol had started to turn sour. The GWR had run into severe financial difficulties, and was having trouble with subsidiary companies and with rivals. In 1846 it had lost the 'gauge war' when the government had ruled that all future railways must be constructed on the standard gauge. The reverberations of the failure of Brunel's 'atmospheric principle' on the South Devon Railway had damaged his reputation in 1848. In the same year, the Bristol Docks Company was acquired by the Corporation of Bristol, and the new Docks Committee began to harry Brunel over his delays in completing the South Entrance Lock. Perhaps the biggest blow of all to Brunel, however, was the accident which befell the S.S. *Great Britain* in November 1846, when the ship ran aground in Dundrum Bay in Northern Ireland. For although the measures taken to protect the ship from the winter storms and to refloat her the following spring were successful, and although her endurance of this ordeal justified confidence in her all-iron construction, the accident brought about the collapse of the Great Western Steamship Company and helped to disperse the precious group of Bristol friends on whom Brunel had relied for good relations with his Bristol clients. So the quality of the relationship changed; formality replacing intimacy, and a chiding tone of anxiety replacing the previous sense of confidence in Brunel's judgment. For the last ten years of his life, until his death in 1859, Brunel had very little to do with Bristol, while Bristol adopted the attitude of disgruntled criticism of Brunel which was still apparent in the works of John Latimer at the end of the century.

Isambard Kingdom Brunel (born April 9th, 1806, died
September 15th, 1859). Portrait by his brother-in-law,
John Calcott Horsley. Bristol City Art Gallery has two
versions.

Clifton Bridge

It is somewhat ironical that the Bristol enterprise of I. K. Brunel which was least affected by the boom in the commercial activity of the city in the 1830s and 1840s and by the subsequent set-backs was the one from which all the others stemmed: the Clifton Bridge. The plan for a bridge across the dramatic Clifton Gorge, where the River Avon cuts its way through a ridge of mountain limestone on its way to the sea at Avonmouth, had been promoted as a result of a legacy of £1,000 left in 1753 by William Vick, a Bristol merchant. The terms of Vick's will were that this money should be allowed to grow by the accumulation of interest until it reached £10,000 when it should be invested in the construction of a bridge across the Gorge. The fund had reached £8,000 by 1829 and, inspired by the recent successful completion of Telford's Menai Suspension Bridge, which seemed to promise a much more economical form of construction for large span bridges than masonry arches, the Society of Merchant Venturers held the competition which fired Brunel's imagination. He applied himself with vigour to the task of designing a suitable structure.

Of the twenty-two designs received by the Merchant Venturers in November 1829, those of Brunel and four others were short-listed. Thomas Telford, the elderly veteran of many major bridge and canal works, first President of the Institution of Civil Engineers, and the outstanding engineer of his day, was called in to judge the competition. He faulted all the schemes submitted. Experience at Menai with lateral wind resistance had convinced him that 600 ft. was the maximum possible span for a suspension structure. As Brunel's designs varied from 870 ft. to 916 ft. for their main span, Telford dismissed them as impracticable despite Brunel's careful engineering calculations to show their feasibility. The result was that Telford was invited to submit a design of his own, which he did, but as the scheme which he presented involved the construction of massive piers from the foot of the Gorge to shorten the central span it was rejected by the Merchant Venturers on account of its estimated cost.

So a fresh competition was held in October 1830, and Brunel submitted a new design in which he compromised with the caution of the Merchant Venturers by reducing the projected span to 630 ft. with a large abutment on the Leigh Woods side of the Gorge. It was this design which was, after some uncertainty on the part of the new judges, accepted in March 1831. Brunel wrote to Benjamin Hawes:

> . . . I have to say that of all the wonderful feats I have performed since I have been in this part of the world, I think yesterday I performed the most wonderful. I

produced unanimity amongst fifteen men who were all quarrelling about the most ticklish subject – taste. The Egyptian thing I brought down was quite extravagantly admired by all and unanimously adopted; and I am directed to make such drawings, lithographs, etc. as I, in my supreme judgment, may deem fit; indeed, they were not only very liberal with their money, but inclined to save themselves much trouble by placing very complete reliance on me . . .[14]

The reference to the 'Egyptian thing' shows that Brunel had responded to the then-current vogue for Egyptian styles in architecture instead of the 'Gothic' style which he had adopted for most of his earlier designs. The conception of the bridge had thus reached a form recognizable as the familiar shape which survives today, even though it underwent much further modification, mainly in order to effect economies.

Having at last agreed on a plan, the Bridge Committee of the Merchant Venturers was anxious to begin work quickly. The first sod was turned on July 21st 1831, to the accompaniment of a ceremony at which Sir Abraham Elton and Lady Elton of Clevedon Court played a leading part. The recorded account of Sir Abraham's address closed on a prophetic note, as he drew the attention of the assembly to the young engineer:

The time will come when, as that gentleman walks along the streets or as he passes from city to city, the cry would be raised: 'There goes the man who reared that stupendous work, the ornament of Bristol and the wonder of the age'.[15]

Little immediate progress was made, however, as there were legal difficulties to be overcome regarding the approaches to the site on the Leigh Woods side of the Gorge, and also it soon became apparent that barely half the £52,000 estimated cost of the project was available, so that new funds would have to be raised. On top of these problems, the Bristol Riots broke out in October, and the whole project was shelved for five years.

A new start was made in 1836. The British Association, founded in 1831, held its sixth annual meeting in Bristol in the summer of that year, and on August 27th the President of the Association, the Marquis of Northampton, laid the foundation stone of the Leigh Woods abutment. Work then proceeded slowly until the two piers were completed in 1840. There was then a pause while a contract was made for the iron work, but in February 1843 it was announced that the sum of £40,000, including the Vick legacy, had been spent and that another £30,000 would be needed to finish the project. This sum was not forthcoming until after Brunel's death, so that the bridge remained for twenty years as merely 'two unsightly piers which deformed the landscape'.[16]

Although work on the Clifton Bridge thus came to a standstill for the remainder of Brunel's life-time, Brunel himself never abandoned hope that the scheme would revive. He was extremely reluctant to accept the decision of the Bridge Trustees to sell off the iron work, which included the two double wrought iron suspension chains, and repeatedly made excuses for not having done so. In the spring of 1849, he wrote to Ward, the Secretary of the Trustees:

> Now a sudden light has come upon me – only I fear to show however our utter darkness – I mean that I really believe that I could get the bridge finished – but what is to be done with the Ferry? If you can devise and procure relief from this I think I will manage the bridge. What can be done – is it hopeless.[17]

From this it is clear that the Trustees were encountering strong opposition from the proprietors of the Rownham Ferry, who operated a flourishing business across the Avon from Hotwells to Rownham Hill. This problem persisted later in the year:

> My Dear Ward,
> Pray do not let anything be done if you can possibly prevent it which can in any degree be a step towards the abandonment of the bridge – out of evil sometimes comes good, and the very badness of the times in all mechanical manufacturing engineering (Engineering properly speaking is not *bad* but *dead*) will be the cause of our finishing the bridge. If I can limit the capital required to manufacturer's capital that is materials and work I can get parties to undertake it, and I think I can manage all (with some little personal sacrifice) except the Ferry.
> If you will get that burden or bug-bear – for it is more alarming than it ever ought to be – removed I think I see my way set to work now manfully, and you will succeed and then I shall be bound to also.
> Yours very truly –
> I. K. Brunel.[18]

The reasons for Brunel's optimism are obscure, but two years later he again tried to make proposals for a resumption of work rather than comply with the Trustees' instruction to dispose of the iron work:

> I cannot think that the Trustees are wise – I am ordered to sell the iron and thus destroy for ever all hopes of completion at the moment when this is at last rendered practicable.[19]

Despite all prevarications, however, the Trustees were firmly resolved to sell the iron in order partially to recoup their losses. Even Brunel's optimism began eventually to disperse. At the end of 1851 he wrote: 'the whole must be closed as I am sick of it'.[20] But his delaying tactics continued, with the desultory negotiations to sell the iron work, until it was finally disposed of to the South Devon Railway Company – for which Brunel was the engineer – which incorporated the chains designed for Clifton in the Royal Albert Bridge over the Tamar, opened in 1859.

Brunel was keenly disappointed by the collapse of this, his first engineering commission, and it may be regarded as symbolic of his relationship with Bristol affairs that the gradual extinction of the Bridge project coincided both with the dispersal of his group of personal friends in Bristol and with an increasing estrangement from other Bristol clients. Yet the Clifton Bridge was completed, but only after Brunel's death and specifically as a memorial to him. A new company was formed in 1861, raising a capital of £35,000 and acquiring the property of the original undertaking. Sir John Hawkshaw and W. H. Barlow were given the engineering responsibility, and made several modifications to Brunel's design. They were able to use the double chains from Brunel's Hungerford Bridge, then being dismantled to make way for Charing Cross Railway Bridge, and they added a new third chain on each side of the bridge platform. Brunel's intended embellishment of the two piers with ornamental cast-iron plates was abandoned. The work was completed in 1864 and the Bridge was formally opened on December 8th that year.[21]

Scene of the laying of the foundation stone on June 21st, 1831 by Sir Abraham Elton of Clevedon Court.
Postcard Lionel Reeves.

Brunel's first great professional success was to win the second Suspension Bridge competition in March, 1831. He managed by brilliant discourse and skilful demonstration with drawings to have his 'Egyptian thing quite extravagantly admired by all and unanimously adopted'.

Lithograph in the Bristol Reference Library from an original drawing in the Swindon Railway Museum.   *Photo John Cornwell.*

Brunel was a competent artist and illustrated each entry design. *Above* One of his designs for the first competition, in 1830.

Below No. 2 design by Brunel for the first competition. Travellers would have emerged over the Avon gorge through a tunnel on the Clifton side. This imaginative and dramatic approach was not considered safe or viable by the judges.
Bristol City Art Gallery. *Photos* John Cornwell.

Bristol Riots: Charge of the Dragoon Guards in Queen
Square. The Riots took place in 1831 and Brunel was
enlisted to be on duty as a special constable when they
were at their height on Sunday, October 30th; he helped
to salvage plate and pictures from the Mansion House. A
year later he gave evidence in defence of the Mayor who
was tried for negligence. The ensuing cost for repairs and
compensation put paid to fund raising for the completion
of Clifton Suspension Bridge.

Engraving from Nicholls and Taylor, *Bristol Past and Present*, Arrowsmith,
1882, Vol III.

The Tower in wooden scaffolding became a favourite background for photographers. By 1843, the abutments were almost completed and then little more was done in Brunel's lifetime. The chains made for the bridge were sold to the South Devon Railway in 1853 for use in Brunel's Royal Albert Bridge at Saltash.

The scene in 1860. Protests in the local press created a threat of demolition, but Brunel's death in 1859 led members of the Institution of Civil Engineers to consider completing the bridge as a monument to its creator.

PIER OF CLIFTON BRIDGE IN 1860.

The Leigh Woods abutment and tower seen from The Colonnade in Hotwells. The Pumphouse and Spa buildings to the left of The Colonnade occupied a site over which the Portway road to Avonmouth was later to be constructed.

Photos Lionel Reeves.

Clifton Suspension Bridge: a view of the piers from the Leigh Woods side.
Photo Reece Winstone.

CLIFTON SUSPENSION BRIDGE IN CONSTRUCTION.

The bridge nearing completion. Work resumed in June 1863 and by May of the following year, the chains from Brunel's Hungerford Bridge were in position.
Postcards Lionel Reeves.

Clifton
Suspension Bridge
nearing Completion

Right Clifton Suspension Bridge: development of the design of the link lugs – a series of drawings by I. K. Brunel.

University of Bristol Library Special Collection. Illustrated in Sir A. Pugsley's *The Works of Isambard Kingdom Brunel*, Institution of Civil Engineers, 1976.

Clifton Suspension Bridge, December 8th, 1864. The
Opening Ceremony involved three processions: first, the
military, followed by the friendly and benefit societies and
trade organisations and finally the gentry. The Lords
Lieutenant of Gloucestershire and Somerset declared the
bridge open and their counties united.

Illustrated London News, December 17th, 1864.

Finished on a simplified modification of Brunel's design by John Hawkshaw and W. H. Barlow, the bridge still gives pleasure and utility to thousands of people every year.

Photo John Trelawny-Ross.

Total length	1352 ft.
Span between piers	702 ft. 3 in.
Width	31 ft.
Height above road	245 ft.
Height of piers	86 ft.
Number of suspension rods	162
Total weight	1500 tons

Dock Works

The Bristol Merchant Venturers who were charmed by the young engineer's advocacy of his Clifton Bridge design in 1831 were responsible for bringing Brunel to the notice of the Bristol Docks Company in the following year. He was introduced to the Board of the company by one of its members, Nicholas Roch, and was invited to make a report on the problems of the docks. These were long-standing and complex. Briefly, the trouble was that Bristol, some eight miles from the sea up a winding and heavily tidal river, was becoming increasingly unsuitable for large ships, at a time when rival ports such as Liverpool were expanding rapidly and overtaking Bristol in commercial prosperity. The problems grew throughout the eighteenth century, and although the eminent engineer William Jessop had greatly improved the dock facilities in 1804–9 by providing permanent high water in the 'Floating Harbour', he had not allowed for a sufficient supply of fresh water to the harbour so that it was prone to silt up and to act as a stagnant sewer for the city. The latter problem had been alleviated by the construction of a culvert in 1828, which redirected the foul water of the River Frome and discharged it directly into the tidal waters of the New Cut. But the silting remained and threatened the continued efficiency of the port of Bristol. It was to the resolution of this problem that Brunel applied his engineering skill.

In his report of August 1832 Brunel suggested various means of improving the flow of water through the Floating Harbour and of removing the banks of mud which were making parts of it unusable.[22] Not all his ideas were immediately adopted, but over the next fifteen years he was frequently consulted by the Bristol Docks Company and commissioned to undertake specific improvements. These included raising the height of the Netham Weir, which directed more water from the River Avon through the Harbour, and constructing the 'trunk' under the dam at the lower end of the Harbour, which converted it from an 'overfall' to an 'underfall' dam and made it possible to scour large quantities of silt into the New Cut. The Company also adopted Brunel's device for a novel form of self-propelled scraper-dredger, one of which was completed in 1844 and remained in active service for over a hundred years before being finally dismantled in the 1960s. A very similar vessel was constructed about the same time for use in Bridgwater Docks, and this has been preserved in Exeter Maritime Museum.

These were substantial improvements, which helped to keep the port of Bristol in business even though they were insufficient to arrest its relative decline compared with Liverpool and other ports. It was clear to Brunel that much more drastic changes were necessary in Bristol, and so when he was asked to give his advice on the condition of the South Entrance Lock he was not inclined to underestimate the seriousness of the problem. He had frequently had cause to warn the Board of the decay of the two entrance locks, and when asked for his opinion on the southern lock (the smaller of the two) in 1844 he observed that: 'the repair of that Lock is a most serious business and will probably involve a *very heavy* expense'.[23] At the same time he wrote to his friend Captain Claxton, who had been closely associated with him in his earlier work on the docks: 'I think of recommending a thoroughly good lock'.[24] The Board of the Docks Company, which was already experiencing severe financial difficulties, was dismayed by Brunel's estimated cost of £22,000, but the need for action was pressing and so it was agreed to adopt his plan for a completely new lock 54 ft. wide and 245 ft. long, which would make it longer than the adjacent North Entrance Lock. In February 1845 the tender of Mr. Rennie, a building contractor apparently unrelated to the famous family of engineers, to do the bulk of the masonry works, was accepted.

Brunel's design for the lock was ingenious, overcoming the limitations of a cramped site and the need to keep traffic moving through the North Lock.[25] The operating length of the lock was maximized by using single-leaf gates of a novel wrought-iron caisson construction so that they became partially buoyant at high water, which made them more manoeuvrable than conventional gates. The work began briskly, but various snags were encountered and the rate of progress slowed down so much that the Board expressed its 'extreme disappointment' in April 1846.[26] Brunel responded promptly and somewhat emotionally in a private letter to the Secretary, objecting to the criticisms of his work:

> My feeling is that I have taken great pains and given great attention as a friend much more than as a professional man to the interests of the Dock Co. as a body and without reference to the contending interests and have amongst other things devised and taken upon myself the responsibility of directing a work which when done will cost about one third what any such work ever did cost – or what any other Engineer would have taken the trouble and responsibility and anxiety of doing for them and have watched their work as I have before watched and carefully directed everything for Bristol and especially for the Dock Co. as amongst my first Bristol clients – and yet I hear nothing but the most contemptible lying reports always in circulation – the most childish impediments thrown in one's way – all by an officer of the Dock Company who by his own weak rather than mischievous character is so entirely beneath my notice as are the reports propagated, that I cannot stoop to wipe him or them out of my way – yet the result is annoying and in any other case I should find some excuse for dropping the business and if the Dock directors really were to ask me any question founded on these reports I believe I should get angry and cut the concern.[27]

A pencilled note against the personal comments on 'an officer of the Dock Company' appears to identify him as Hillhouse, who was the Dock Master for the BDC, but it has not been possible to establish the nature of the criticisms or the role of Hillhouse in giving currency to them. Brunel went on in the same letter to assert his complete confidence in the contractor and in the work so far completed.

This outburst does not seem to have achieved any recovery in confidence in the project on the part of the BDC, and there was certainly no discernible increase in the progress of work on the new lock. Indeed, there was a serious set-back a few months later when a retaining dam collapsed in the lock, causing further long delays. Brunel had his Chief Assistant, J. W. Hammond, engaged on the Bristol Docks work, and it was an additional blow to Brunel when Hammond died suddenly the following year. Then relationships were further complicated in September 1848, by the winding up of the impoverished BDC and the assumption of responsibility for the docks by the Corporation of Bristol, which established a Docks Committee to exercise control. One of the first resolutions of this Committee called on Brunel 'to adopt every means in his power to place the Lock in working order as soon as possible'.[28] But there were more delays before the lock gates were in working order (probably about April 1849) and the swing bridge over the lock, with its interesting wrought iron tubular girders anticipating Brunel's later major works in this form of construction at Chepstow and Saltash, was not completed until the following autumn.[29]

Meanwhile, Brunel wrote another prickly letter to Osborne on a related matter:

> I have heard accidentally perhaps incorrectly that at the last Dock meeting instead of replying to my inquiries some question arose as to them consulting me *at all* on the subject of the steam boat landing places. Now if any such question was asked I must beg of you to put the matter quite clear. It was no wish of *mine* but I was specifically asked by the Directors or I suppose more correctly by members of the Committee to consider the subject and give them some plan. . . . I only require to be told what is expected I can then choose for myself.[30]

Brunel had long been interested in improving the facilities for steam ships in the Bristol Docks. His own S.S. *Great Western* had difficulty in using them, and the tremendous problem encountered in getting the S.S. *Great Britain* out of the Floating Harbour was one of the most powerful incentives to improving the South Entrance Lock. But no progress was made during Brunel's lifetime towards equipping the port of Bristol to cope with the large steam ships which were coming into service in the 1840s and after.

Mutual dissatisfaction between Brunel and the Dock authorities had thus become well established by 1848, and it appears to have been a relief to both parties when the end of the relationship came in sight. However, it was not to come smoothly, because there were first some accounts to be settled, and a long wrangle ensued about the entitlement of the contractor Rennie to claim payments additional to his contract for sums incurred as a result of verbal instructions from the late Mr. Hammond, which had not been ratified on paper. Rennie claimed a sum of £2759.8.0d, and after due consideration Brunel supported him up to

£2000.[31] The Committee responded with an offer of £1000 which Rennie refused, but a compromise was eventually reached by the end of 1852.[32]

This settlement was the effective termination of Brunel's active involvement in Bristol Docks. Apart from a few points of advice on minor engineering details, the only subsequent business Brunel had with the Docks Committee was in 1858, when Bristol was once more considering major port improvements, and he was approached about the possibility of undertaking them. He was out of the country at the time, seeking recuperation from the long illness which was already afflicting him, but on his return his Clerk wrote to L. Bruton of the Bristol Docks Committee:

> Mr Brunel has returned to England and I have communicated to him the movement that is being made at Bristol towards improving the Port. . . . I am writing this note to you as a *private* one by Mr Brunel's desire to say how much gratified he would be if the Bristol people were to put the matter into his hands, so many of them being his friends, the Miles', Brights' and others, and the circumstance that it was at Bristol that he commenced his professional life. At the same time Mr Brunel would wish it to be clearly understood that he would not on any account wish that the promoters of the intended projects should on his account be prevented from consulting and employing any other Engineer, rather than himself, if by so doing all interests in the work would be more benefitted, and in any other case, Mr Brunel would feel much disappointed if he had not the offer made to him to become the Engineer of so interesting and important a work.[33]

Nothing came of this approach, and a little over a year later Brunel was dead.

Bristol had to wait another decade for further port improvements, and then on a much more modest scale than Brunel would have projected. The fact was that the traditional port of Bristol, in the heart of the city several miles from the sea, was no longer suitable for ocean traffic. Brunel had realised this as early as 1844, when submitting his proposals for the reconstruction of the South Lock:

> I have recommended these dimensions because I believe they would be sufficient to accommodate all ordinary Steam Boats built for the Irish Trade – and this I now think is sufficient for the Port of Bristol.[34]

The future for the port lay nearer the mouth of the Avon, and, as we will see, Brunel had views on this also. But the fact that the Floating Harbour survives today is in part a monument to Brunel, for it incorporates his improvements in water flow and scouring techniques which have prevented it from silting up. His South Entrance Lock, although long disused and deprived of its spectacular wrought-iron caisson gates, is still an impressive piece of masonry when viewed at low tide, when the recess for the outer gate is fully visible together with the lip in the stonework against which it closed and the channels in the masonry through which ran the chains to the capstans. And the swing bridge which Brunel designed to cross the lock is still intact but disused alongside the North Entrance Lock, its place over Brunel's Lock having been taken by a substitute constructed on similar lines and now fixed in position.

Above **Brunel's** lock, with its beautiful elliptical arch shape can be seen on the right of this 1866 photograph. Over it swung the wrought iron tubular girder bridge

Photos **P.B.A.**

Below At the request of the Bristol Docks Company, Brunel submitted a report making various recommendations for improvements. One was the use of a drag boat which would pull itself across the harbour on chains attached to bollards. The B.D.6 was built to Brunel's specification in 1843 and worked until 1961. The pole at the prow of the boat is the head of a scraper which stirred up mud to be flushed out of the harbour through conduits.

Above South Entrance Lock at Cumberland Basin. Brunel's lock is 262 ft. long and 54 ft. wide. A concrete wall now replaces the original floating gates and it is crossed by a replica of the original bridge.

Below Brunel's original wrought iron girder bridge which now lies along the North Lock under the Cumberland Bridge. This was the engineer's first use of tubes made of iron plates, a method he repeated on the Chepstow Railway Bridge and the Royal Albert Bridge, Saltash. *Photos* John Cornwell.

Black Rock Pumping Station, Avon Gorge, 1845. Brunel designed this fantastic engine house to pump water from a large spring to supply water for Clifton. The Society of Merchant Venturers, patrons of the scheme, were bought out by the newly formed Bristol Water Company.

Photo Bristol City Museum.

Railway Works

Although his life span was so short, Brunel was one of those rare men to become legends in their own time, and the outstanding reasons for his prominent place in the public imagination were his identification with the visionary Great Western Railway and the three great ships which in a sense stemmed from it. Latimer, of course, was correct in judging the adoption of Brunel's schemes such as the broad gauge as economic mistakes, and if the counsels of caution and economy had prevailed over vision and imagination Brunel's services would have been anathema to all railway companies. Latimer quotes with approval the remark by George Stephenson that all the railway lines in the kingdom would eventually be linked together as an overwhelming argument for the narrow gauge.[35] But Brunel's objective on the Great Western Railway was to provide a radically new system of comfortable high-speed travel, and if the detailed fulfilment of this aim could have matched its imaginative conception it is more than likely that all main lines would have adopted his broad gauge. The remarkable thing, after all, is not that Brunel failed but that he came so close to success.

The broad gauge system of 7 ft. rather than the standard or narrow gauge 4 ft. 8½ ins. was established first between London and Bristol. It was then pushed on into Devon and Cornwall and fanned out into South Wales and the Midlands. From his appointment as Engineer to what was shortly to become the GWR in 1833 until his death Brunel master-minded one of the largest railway empires in the country, and the only one which attempted to apply rational and systematic principles to the task of developing a new pattern of high-speed transport.

The new railway was promoted by a group of the same enterprising Bristol merchants and businessmen who supported the Clifton Bridge project, harbour improvements, and other innovations in the 1830s. They included Robert Bright, John Cave, Henry Bush, C. P. Fripp, Peter Maze, George Gibbs, John Harford, and T. R. Guppy, by most of whom Brunel was known and admired, and in the case of Guppy at least he was shortly to become a close personal friend. It is hardly surprising, therefore, that when Nicholas Roch suggested him as their engineer they were ready to accept him, even though he refused to agree with their proposal to adopt the lowest estimate for the route. Brunel warned them:

> You are holding out a premium to the man who will make you the most flattering promises, and it is quite obvious that he who has the least reputation at stake, or the most to gain by temporary success, and least to lose by the consequence of a disappointment, must be the winner in such a race.[36]

The selection committee accepted this admonition from the 27-year-old engineer and appointed him. Both these Bristol men and the London men who joined them – George Henry Gibbs, Charles Russell, C. A. Saunders, and others – were charmed and impressed by Brunel's abilities, and remained strikingly loyal to him throughout the early vicissitudes of the GWR. Even though the original preponderance of Bristol men dwindled over the years as the GWR system expanded, Brunel's engineering leadership was never seriously challenged from inside the company.

The Great Western Railway Act was approved, on the second attempt, in August 1835. Five years later the stretch from Bristol to Bath was opened to public traffic, and by June 1841 rail communication was completed from Bristol to London. In the same month, the Bristol & Exeter Railway, also under Brunel's engineering control, was opened from Bristol to Bridgwater. Exeter was reached by 1844, and the South Devon Railway – where Brunel experimented with the ill-fated atmospheric system – carried the broad gauge through to Plymouth by 1848. Meanwhile, the Bristol & Gloucester Railway, authorised in 1839, was opened as a broad gauge railway in 1844, although it was subsequently acquired by the Midland Railway and converted to standard gauge. The Wilts, Somerset & Weymouth Railway was authorised in 1845, and many other important links put in hand under Brunel's engineering direction, one of the last and most significant being the South Wales Junction project which was still incomplete at the time of his death. At the centre of this Bristol network was the GWR terminus at Temple Meads, designed by Brunel and opened in 1841, and the B & E terminus built at right-angles to it – an arrangement which drew further caustic comment from Latimer.

The main features of the GWR and its associated railways have been very adequately chronicled, and it is not necessary to review any of the details here.[37] For our purposes it is sufficient to observe that the establishment of this great railway empire was one of I. K. Brunel's outstanding achievements, and that it was built largely on the base of Bristol enterprise and Bristol capital. Other interests, and particularly those of the London promoters of the GWR, were represented in the managerial direction (although care was taken to exclude the dissident voices of Liverpool and other northern shareholders), but Bristol remained the natural hub of the broad gauge system. Many of the regular company meetings were held there, and much of the business in the early years was administered from Temple Meads.

The directors confronted an especially difficult meeting of proprietors at Temple Meads station in August 1849, when they proposed reducing the dividend from 4 per cent to 2 per cent. The circumstances of this half-yearly meeting were not propitious, with the railway boom clearly running itself out, with the collapse of George Hudson's Midland Railway empire, and with the recent engineering problems on the South Devon Railway, where Brunel had been obliged to recommend the abandonment of the atmospheric experiment. In addition, a serious cholera epidemic was then raging in the city of Bristol. But the persuasive oratory of Charles Russell and C. A. Saunders, the Chairman and Secretary of the GWR respectively, succeeded in mollifying the critics. The local newspaper reported that:

> Little Isidore (*sic*) Brunel, creeping into a corner, laughed securely over Lord Barrington's shoulder.[38]

But the same newspaper recalled this incident twenty-two years later in less disparaging terms, when giving a notice to the publication of the biography of Brunel by his son. Under the heading: 'Isambard Kingdom Brunel and the Old Railway Days' it commented:

> Though the 'little giant', as someone called him has been 'gone from our gaze' about thirteen or fourteen years, we can still see him with our mind's eye sitting 'calm as the halcyon' behind the row of Directors at the stormy half-yearly meetings in Temple-meads. . . . He contemplated the hurly-burly before him with the composure of a philosopher and the quiet enjoyment of a humourist, even though his own 'extravagance' was the subject under discussion.[39]

Brunel's presence was certainly regarded as essential to the smooth running of any GWR meeting in those formative years.

Brunel's relationships with the other Bristol railways were not as smooth as those with the GWR. He engineered the line of the Bristol & Exeter Railway, opened through to Exeter in 1844. But in 1846 he had a disagreement with the Board about differences with the GWR, which he felt made it impossible to remain as Engineer to both. So he resigned from the B & E observing in his letter of resignation to James Gibbs: 'I am hardly able to recognise my former Directors in the proceedings of the present Board'.[40] The Bristol & Gloucester line also opened in 1844, but in 1846 it was acquired by the Midland Railway so that Brunel was deprived of this engineering responsibility. He had previously advised the directors against the construction of a new goods station:

> Bristol is not Liverpool and if you carried for nothing you would not make a larger *regular Trade* – than even at moderate prices. . . . Railway people at the present moment are all mad – and excuse my saying it the B & G amongst the number but when Nov comes – and shares drop and the public begin to learn that you and the rest have been carrying on an enormous trade at a loss like the cheap linen Drapers there will be a dreadful reaction.[41]

On the southern branches from Bristol, Brunel responded to an invitation in May 1844 to become engineer for the Weymouth line, but expressed reservations about the proposed route:

I observe you speak of a *Bath* and Weymouth line – now I think the practicability of a direct line from Bath itself is more than questionable. The Gt. Westn. Rly Comy are now projecting a line to Frome. . . . On the other hand it is quite necessary to consider whether the Bristol & Exeter Ry does not really offer the most advantageous line of communication.[42]

In the last year of his life, Brunel was engaged on the Bristol & South Wales Junction Railway, projecting a line northwards from the city through Ashley Down and Patchway to the proposed New Passage Ferry, where trains ran out onto timber piers from which the passengers transferred to the waiting boats. He was concerned to obtain the services of a competent Resident Engineer to do the preliminary work and wrote to one of his former Assistants, C. Richardson:

I want a man acquainted with tunnelling and who will with a moderate amount of inspecting assistance look after the Tunnel with his own eyes – for I am beginning to be sick of Inspectors who see nothing – and resident engineers who reside at home and he must be one to whom salary is not the principal object.

It is interesting to find Brunel writing with such vehemence, arising perhaps from his Herculean labours at the time on the S.S. *Great Eastern*, where he undoubtedly felt let-down by those on whose work he relied. As far as Richardson was concerned, however, Brunel was only able to offer a salary of £300 p.a. rising to £450 when the full works commenced, but in compensation he pointed out that:

the country immediately north of Bristol I should think a delightful one to live in – beautiful country – good society near Bristol & Clifton etc. – I can't vouch for any cricketing but should think it highly probable.[43]

The offer of cricket as bait to Richardson is curious, because five years earlier Brunel had reprimanded his Assistant for over-indulgence in this exercise.[44] Even with this temptation, however, Richardson did not find the salary attractive, but as he was not then in employment Brunel was able to press him:

If you desire on the other hand to go to work again. . . . I think this an excellent opportunity and again offer it to you provided you assure me that you enter upon it with pleasure and that a resumption of active life is an object to you.[45]

Richardson accepted and Brunel soon found a pretext for increasing his salary to £400. The project was delayed by Brunel's death, but was eventually completed in 1863.

The extension of the broad gauge network of railway lines from Bristol was a matter of great concern to Brunel, and until the Midland Railway acquired the Bristol & Gloucester and pushed the narrow gauge through to Temple Meads in 1854 he had a virtual monopoly of railway engineering in the region. Many physical monuments remain of this activity, from the railway routes themselves to many points of detailed design such as the decorative entrances to the tunnels between Bristol and Bath. The most important of these monuments, however, is the original Temple Meads terminus, still substantially intact although converted into a car park and in a poor state of repair. Brunel designed the station with the wide wooden-beamed roof as it exists today except for the introduction of a channel for light and ventilation along the ridge. This gives an unimpeded space 72 ft. wide between the two rows of iron columns which carry the timber beams in cantilever fashion, with their outermost extremities anchored in the masonry walls. The elegant 'hammer beams' were added for decorative effect and have no constructional significance. This train shed is linked to an impressive office building with a neo-Tudor façade facing onto Temple Gate, which is intact except for one wing which was removed to make way for the approach ramp to the enlarged Temple Meads station in 1878. The whole structure survives as the oldest railway terminus in the world to preserve most of its original shape, and as such it is a national monument of outstanding importance. This has been recognized by designating it a Grade I Listed building, and in October 1981 British Rail leased the site to a charitable trust, the Brunel Engineering Centre Trust, which has undertaken to restore the building and to refurbish it as a heritage feature.[46]

Young Brunel at the beginning of his career. Before him is spread his survey for the Great Western Railway. Portrait by John Horsley. *Photo* Public Record Office.

Above South East view of Brunel's railway terminus. The ticket office at the front had three floors and attic accommodation with a Board Room on the first floor. The façade bears a striking resemblance to R. S. Pope's Bristol Guild Hall in Broad Street.

Below The main façade. Passengers arrived by coach or on foot and entered the station through the arches on the left. A difficult ascending staircase had to be negotiated to reach the platform above the basement level. Traffic passed under the line to make its exit from the right hand portal.
Bristol City Museum.

Temple Meads Station left hand portal. Notice the clock which enabled passengers to check on punctuality.
Bourne lithograph.

The Temple Meads Board Room, still intact; the scene of many a stormy directors' and shareholders' meeting during the early days of the Great Western Railway.
Photo Christopher Dalton.

This lithograph by J. C. Bourne (1842) captures the beautiful spatial grandeur of this earliest surviving railway terminus. The mock hammerbeam roof, the arcades of Tudor arches and the broad gauge lines, engines and carriages evoke the glamour and excitement of rail travel in the early days.

The colonnade on the station platform.
Vaughan, *Great Western Architecture*, Oxford Publishing Co., 1977.

Temple Meads Station. The northern end of Brunel's
engine shed is on the right. In the top left is the grand
Jacobean-style Bristol and Exeter Railway office of 1852.
Photo John Cornwell.

The original Temple Meads goods shed. Bourne lithograph.

Left Daniel Gooch (1816–1889). Chosen by Brunel at the age of 21 to be his locomotive superintendent of the Great Western Railway, he built the best of the broad gauge engines and planned the company's railway works at Swindon. In 1860, Gooch was appointed Engineer to the PSS *Great Eastern* and converted her to lay the Atlantic cable, the most useful part of her life.

Below The North Star, Daniel Gooch's locomotive, designed in 1836, was originally built for export to Russia for a 6 ft. gauge line but adapted for Brunel's Great Western Railway. She was the first of a line of Broad Gauge Flyers.
Photo John Cornwell.

Above Today, this bridge over the Avon near Temple Meads is almost hidden, but also protected, by a steel bridge constructed to carry extra lines. Between Bristol and Bath the GWR was characterised by fanciful and graceful Tudor or Gothic works. Brunel's son recorded that his father 'took great pains in finishing minutely the various designs, making them correct in their proportions and details'.

Below Another Bourne lithograph, of Tunnel No. 2 near Bristol. A romantic ruined gateway: the ground to the left having slipped away during construction, Brunel decided to leave it unfinished and plant it with ivy.

Sydney Gardens, Bath. An illustration of Brunel's tasteful insertion of his railway into a fashionable recreational area. The graceful retaining wall, balustrades and footbridges create a setting against which to admire the passing locomotives.
Bourne lithograph.

The West front of Box Tunnel. In Bath and just beyond
it, Brunel built in an Italianate classical style. The
legend that the sun shines through the tunnel on the
morning of Brunel's birthday has been disproved.
Bourne lithograph.

Above **The Gover Viaduct**. This wooden viaduct near St. Austell, like the many others constructed in Cornwall and South Devon, was built with standard spars and beams for easy replacement. It was replaced with stone arches in 1898.

Vaughan, *Great Western Architecture*.

Below The last broad guage train leaving Paddington. On the morning of May 20th, 1892, thirty-three years after Brunel's death, the battle for the 7 ft. guage was lost. Conversion to the standard 4 ft. $8\frac{1}{2}$ in. wide track between Paddington and Penzance was completed in 48 hours.

Photo British Rail.

The Bristol Ships

The story has frequently been told of how Brunel astonished and delighted members of the GWR board of directors by suggesting that their railway could be extended from Bristol to New York by way of a steam ship. Although it is not certain whether the initial conception of this idea was due to Brunel or to Guppy, the directors gave it their approval and Brunel was presented with the opportunity to reveal his genius as a marine engineer. There is a saga associated with all three of Brunel's great steam ships, but only the first two were built in Bristol and concern us here. The third, the S.S. *Great Eastern*, was built in London and was so large that there were few ports in the world with facilities to handle her, and there was certainly never any question of her visiting Bristol.

While supporting the enterprise into ship building, the directors of the GWR considered it prudent to form a new company for the purpose, and the Great Western Steamship Company was duly set up in 1835. Peter Maze, a Bristol merchant with interests in railways and the cotton industry (his large cotton mill, opened in Barton Mill in 1838, was predictably called the Great Western Cotton Mill), was appointed chairman, and Christopher Claxton was made managing director. The first ship, the S.S. *Great Western*, was a timber-hulled paddle steamer designed by Brunel and built by William Patterson, a prominent Bristol shipbuilder, in his yard at what is now Prince's Wharf, opposite the point where the River Frome joined the River Avon in the middle of the Floating Harbour. The London engineer Henry Maudslay provided the power unit with a pair of side-lever steam engines. The ship was launched on July 19th 1837 and made its maiden voyage to New York in April 1838, being narrowly beaten for the honour of being the first westward steam-powered crossing of the Atlantic by the *Sirius*, an Irish steam ferry hastily adapted for the purpose by Liverpool merchants who had become anxious at the spectacular burst of energy of their old rivals. The *Great Western* went on to enjoy a remarkably successful career which did much to convince the world of the feasibility and reliability of steam ships in ocean service. By doing so, she upset the dire prognostications of Brunel's critics, who had predicted the failure of the enterprise, and established the reputation of her designer as a marine engineer.[47]

The logistics of the north Atlantic trade required that the *Great Western* should have a sister ship, so that the two vessels could maintain a regular shuttle service of passengers, goods, and, it was hoped, mail. Thus the Great Western Steamship Company began work in July 1839 on the construction of a second ship. But Brunel had by now become fired by the idea for a much larger vessel, for which he quickly adopted iron as the material for the hull and went on to abandon the design for paddle wheels in favour of screw propulsion. So was born the S.S. *Great Britain*. As Patterson's yard was too small to contain the new ship, Brunel enlarged an existing dry dock on the south side of the Floating Harbour and built the ship in the dock. Not only did he have the major responsibility for the hull, which was – and is – a brilliant work of technical perfection, but he also designed the engine and the screw. All of this required detailed experiments, consultation, and reports, of which only fragmentary documentary evidence survives. Brunel worked closely with Claxton and Guppy, the other two members of the Building Committee to whom the directors delegated the work of construction, and on July 19th 1843 the ship was ready to be floated out of her dock at a ceremony performed by HRH Prince Albert.

Unfortunately, the ship was then trapped in the Floating Harbour for a further eighteen months, because the Cumberland Basin locks through which she was obliged to pass to reach the sea were too narrow to accommodate her. This was before Brunel had undertaken the widening of the South Entrance Lock, and involved delicate negotiations with the Docks Company, which had already adopted an unhelpful attitude towards the *Great Western*, the paddle steamer having been refused a decrease in harbour dues on account of its inability to use the main harbour facilities regularly. While thus marooned in the Floating Harbour, the *Great Britain* was photographed alongside Mardyke Wharf, Hotwells, by the local photographic pioneer Fox Talbot, making this one of the earliest industrial archaeological photographs. Brunel eventually received permission to modify the lock-fittings sufficiently to squeeze the ship through the locks at Cumberland Basin, and the *Great Britain* was able to leave the port of Bristol on the highest available tide, and even then only with great difficulty, on December 11th 1844. Late that night Brunel wrote to the directors of the South Wales Railway apologising for his inability to attend their meeting the next day:

> We have had an unexpected difficulty with the Gt. Britain this morning – she stuck in the Lock – we *did* get her back – I have been hard at work all day altering the masonry of the lock – tonight our last tide we have succeeded in getting her through but being dark we have been obliged to ground her outside – and I confess I cannot leave her till I see her afloat again and all clear of our difficulty here – I have as you will admit much at stake here and I am too anxious to leave her.[48]

Once she was away from the Floating Harbour, the link between Bristol and the *Great Britain* was effectively broken – at least until her triumphant return from the Falkland Islands in the summer of 1970, when she was floated back into the dry dock which had been specially designed by Brunel for her construction. This was on July 19th, thus commemorating the launching day of both herself and the *Great Western*. The older ship had already started to use Liverpool regularly in 1843, and when the Great Western Steamship Company collapsed in 1847, following the stranding of the *Great Britain* in Dundrum Bay and the expensive business of re-floating her, the *Great Western* was sold at an auction and spent the next nine years operating out of Southampton until it completed its useful life and was scrapped in 1856. From the end of 1844, therefore, Brunel's ships had nothing more to do with Bristol.[49]

Brunel's first ship. *The launching of the Great Western, 1837,* an impression by
the marine painter Arthur Wilde Parsons. The ship was launched from a
dock near the Prince Street swing bridge and the Bristol Industrial
Museum. A plaque records the event.
City Art Gallery: the painting hangs in the Old Council House, Corn Street.

William Patterson (1795–1869). A friend and colleague
of Brunel's, he did much of the supervision of the
building of the *Great Western* in his own Wapping Yard.
He drew the 'lines' of the *Great Britain* before she was
built, and with Brunel and Claxton collaborated in the
first salvage of the ship when she was beached in
Northern Ireland.
Photo Bristol City Museum.

The Royal Western Hotel, designed by R. S. Pope in collaboration with Brunel was erected between 1837 and 1839 to accommodate travellers going from London to New York by steam. After a train ride from Paddington and a night spent in the hotel, passengers embarked in the *Great Western* for America. The building ceased to be a hotel in 1855 and functioned for a time as a Turkish Baths. Today, the building now houses local council offices.

Postcard Lionel Reeves.

The *Great Western*, built of oak reinforced with iron in William Patterson's yard. She was launched on July 19th, 1837.

Overall length	236 ft.
Breadth	59.6 ft.
Capacity	148 passengers

Bristol City Museum. *Photo* John Cornwell.

The launch of the SS *Great Britain* on July 19th, 1843. Here we see her floating out of the dock specially constructed for her and the place to which she returned 127 years later. In the illustration she floats high in the water because her engines had not been fitted. She was the first ocean-going ship to be built of iron and driven by screw propeller.
Bristol City Museum.

Overall length	322 ft.
Breadth	51 ft.
Capacity	360 passengers

Poster advertising seats to witness the historic event.

Photo Bristol City Museum.

Above SS *Great Britain*: this historic photograph was taken by William Fox-Talbot from the Great Western Dock looking across to the Northern Gas Ferry steps. This is thought to be the first photograph to use a ship as a subject.
National Maritime Museum.

Below The *Great Britain* went aground in Dundrum Bay on September 22nd, 1846, the ship having earlier set out from Liverpool on her fifth voyage. This calamity put an end to the Great Western Steamship Company and could have finished the ship. She was salvaged by being protected from the winter storms by a mattress-like breakwater, and was refloated the following spring.
Watercolour, J. Walker, 1847. Science Museum, London.

Section at After end of Boilers.

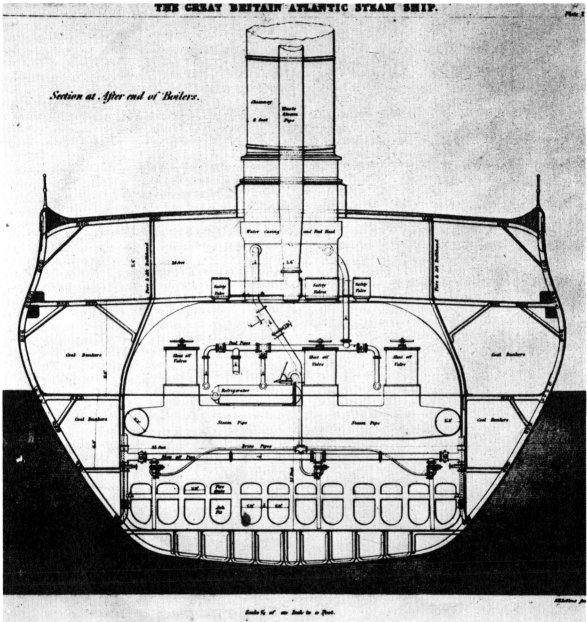

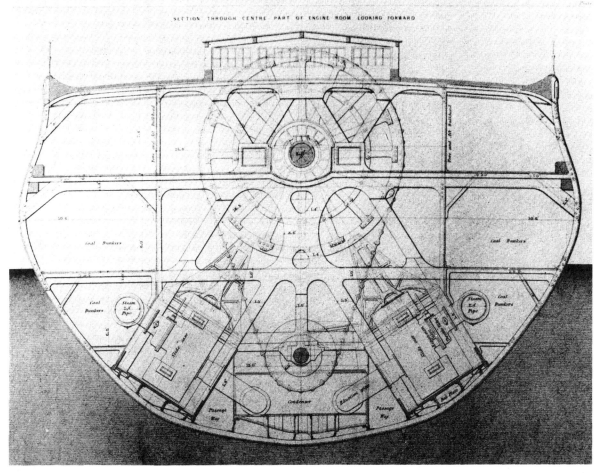

SECTION THROUGH CENTRE PART OF ENGINE ROOM LOOKING FORWARD

Left This drawing (showing section at after end of boilers), and those on the following pages, is taken from a book of 25 drawings published by John Weale of Holborn.

Above Section through centre part of the engine room looking forward. The dominating feature is the huge spoked wheel and chain drive to the propeller shaft to increase the speed of the screw.

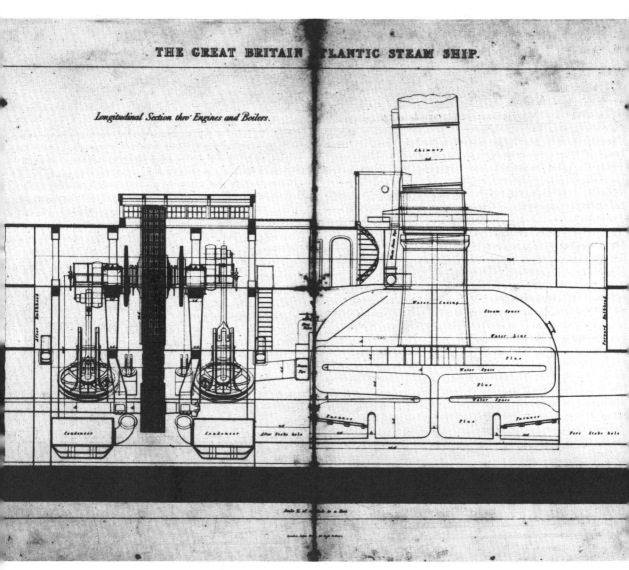

Longitudinal section through the engines and boilers.

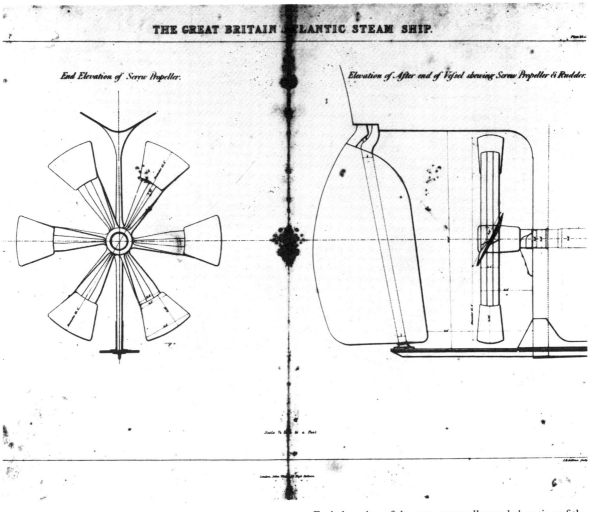

End Elevation of Screw Propeller.

Elevation of After end of Vessel shewing Screw Propeller & Rudder.

End elevation of the screw propeller and elevation of the after end of the vessel showing the screw propeller and the rudder. The six-bladed screw had an overall diameter of 15 ft. 6 in. and weighed 3 tons 17 cwt. The rudder was balanced and streamlined to suit screw propulsion. It was 14 ft. high and 7 ft. 6 in. in mean width.

Above The *Great Britain* in 1852 making her first voyage to Australia. The ship underwent a number of alterations during her working life. Here, her six original masts have been reduced to four and her single funnel is coupled with another.
Painting – *Dropping the Pilot at Liverpool Bar*, 1852. National Maritime Museum.

Right Beached in Sparrow Cove, 1960.
Photo John Poltock.

In the Falklands. *Left* The main mast, 1960. *Below* On the deck looking towards the stern. The ship deteriorated a great deal in the two years between this photograph and the beginning of her final salvage.
Photos John Poltock.

The *Great Britain* in Sparrow Cove waiting to be salvaged. The ship began her 47th and last voyage on February 6th, 1886, to carry a cargo of coal from Penarth to San Francisco. A storm caused some damage and she pulled in to the Falklands for repair and stayed there as a storage hulk until 1970.

The return: on her long journey home the ship travelled high and dry on the raft *Mulus III*. The pontoon was 250 ft. long and 79 ft. wide; it was sunk beneath the great ship and then filled with air until it floated, lifting the vessel out of the water. The *Great Britain* left Port Stanley on April 24th, 1970, and arrived at Avonmouth, Bristol, three months later.

Photos Great Britain Project.

The *Great Britain* in Jeffrey's Dock, Avonmouth, July 2nd, 1970.

The *Great Britain* in 1981, in her final home in the city of Bristol.
Photo SWPA.

The *Great Eastern*: Brunel's last and most ambitious ship, driven by paddle and screw, was large enough to carry fuel for a return journey to Australia. There were many problems encountered during her construction, and these were not a little aggravated by the differences between Brunel and his collaborator John Scott Russell, a man whose knowledge of ship design and construction was at least as great as Brunel's. This great ship was launched broadside at Millwall in London in 1858.

This actual lithograph was the property of Brunel's Chief Assistant, R. P. Brereton.

Overall length	692 ft.
Width over paddles	118 ft.
Capacity	4,000 passengers

Left A high pressure water jet blasting off decades of accumulated barnacles and rust.
Photo Plant Technical Services Ltd.

Above Restoration in progress: the prow, eroded by rust, being prepared for the fibre-glass infill.
Photo Newton Freeman.

Below H.R.H. Prince Philip in the hold of the *Great Britain* with Richard Goold-Adams (Chairman of the *Great Britain* Project) and Dr Ewan Corlett (the distinguished naval architect in charge of the salvage) inspecting the bottom of the hull.
Photo 'Bristol Evening Post'.

Left The restored deck looking forward. *Photo* SWPA.

Above The stern restored with decorative mouldings and the Arms of the City of Bristol. At the bottom right of the picture is the replica of the six-blade propeller. *Photo* SWPA.

Below The unicorn being restored to the port side of the prow. *Photo* SWPA.

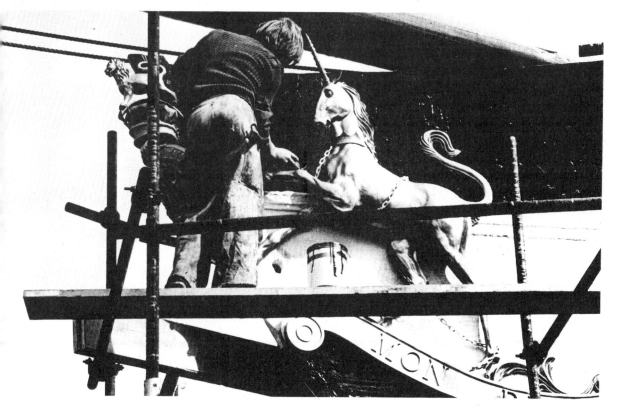

Left Arrival in Bristol July 4th, 1970. On the last lap of her voyage, the *Great Britain* has passed under Brunel's Suspension Bridge. When she made her only journey down the River Avon 127 years before, the construction of the bridge had only just begun.
Photo SWPA.

Below The SS *Great Britain* floating into the Great Western Dock. Her birthplace became her final resting place on the evening of July 19th, 1970.
Photo Christopher Burnell.

The *Great Eastern* at Dublin. Having successfully laid the Atlantic Cable, the ship suffered the ignominy of being used as a fun fair and floating exhibition. Her visit to Ireland for this purpose was commercially disastrous and she was then coaxed up the Mersey to Liverpool, Bristol's rival port, to be scrapped.

National Library of Ireland: Lawrence Collection.

Brunel standing by the checking drums whose chains were used to control the launch of the *Great Eastern*. The most widely known picture of Brunel was most likely taken at the same time. This photograph bears his signature.

From Lady Celia Noble *The Brunels, father and son*, Cobden Sanderson, 1938.

Waterworks

The decades of the 1840s and 1850s were distinguished by a new civic consciousness in Britain, arising out of the cholera epidemics and other scourges, which showed itself amongst other ways in the establishment of pure water supplies. Brunel was consulted on several such water supply schemes, including those designed for Glasgow by J. Latrobe Bateman, so it was appropriate that he should make a contribution towards the health of Bristol in this way. He was consulted by the Bristol Society of Merchant Venturers in 1841 with a view to setting up a 'Clifton Water Works' scheme, and the following year he reported to the Committee of the Society that the estimated cost of establishing water works at Clifton would be £1,364.[50] Brunel proposed to tap a spring discovered in the Clifton Gorge in 1836, and was optimistic about the amount of water which it would provide. A pumping station was duly built at Black Rock in 1845, its 'somewhat fantastic design' winning a sneer from Latimer.[51] Surviving photographs of the building show a structural similarity with the Italianate engine houses designed by Brunel for the atmospheric section of the South Devon Railway.

In 1846, Brunel was called in again by the Society of Merchant Venturers to consult on their behalf with the representatives of a rival undertaking, the Bristol Water Works Company. This had been set up by Dr. William Budd, the pioneer epidemologist, and other citizens to supply the whole of the city rather than just Clifton, and their engineer James Simpson, a nationally renowned expert in the field, had advised drawing on the abundant water from the Mendip foothills to the south of Bristol. Brunel duly wrote to the rival company:

> The promoters of the Clifton Water Works Bill will be very happy to meet any parties representing the promoters of your water works Bill to endeavour to see if a more satisfactory plan can be suggested by which the public objects they have in view can be attained, and which at the same time shall be satisfactory to your company – It appears to them that a fair Division of district will be the most rational principle on which to go, but this may be a matter of discussion.[52]

But Parliament preferred the more comprehensive scheme, and this eventually absorbed the Society's water works, which received £18,000 in compensation for its plant. The Black Rock pumping station was demolished in 1864 to make way for the railway through the Clifton Gorge to Avonmouth. Latimer commented that it was 'a puzzle to strangers owing to its bizarre architecture' and said that it was considered converting the building into a church for sailors and bargemen.[53]

78

Brunel's professional relationship with Simpson remained very cordial throughout the negotiations about the rival water works. He wrote to him in 1848 regarding the possibility of using hydraulic power at his installations on the South Entrance Lock:

> My Dear Sir – I would very much like to use some water power at Bristol at two different places – Temple Meads and near the Locks at Cumberland Basin. Will you tell me whether you have now or are likely soon to have and *when* water laid on with a high pressure and *what pressure* at either of these places? Some plans of mine are entirely dependent on your answer, and I shall therefore be glad to hear at your earliest convenience.[54]

Brunel was experimenting with hydraulic machinery at Temple Meads for operating lifts in order to connect the goods shed with the main line, 12 ft. above it. Such apparatus was still in its infancy and was not used at the entrance locks until the next major improvements were made there, twenty years after Brunel's death.[55]

Miscellaneous Works

We have now reviewed the major areas of Brunel's activities in the Bristol region, but there were a number of minor items of business which, although of less significance in terms of their achievement, deserve a mention. It is not generally realised, for instance, that Brunel was consulted about structural work at Bristol Cathedral. The Dean invited him to examine some defects in the fabric of the Cathedral in the summer of 1850, so that Brunel replied to him proposing:

> to give a couple of hours for a cursory inspection to form some opinion of the subject. . . . I can be at the Cathedral at 5 o'clock on Tuesday morning next, having to leave Bristol by train to Exeter at 7.50. . . .[56]

This early appointment must have been kept, as three weeks later Brunel submitted his report to the Dean. He referred favourably to a previous survey by J. F. Welsh, but differed from this in thinking that the defects were those to be expected in any old building and that they presented no immediate threat:

> Mr. Welsh no doubt would feel little anxiety as to the building standing 24 hours – I feel little as to its standing 24 years – we both feel that repairs are required. . . . The next or the following generation will probably be called upon for more extensive repairs and restoration.[57]

Despite this reassurance, the Dean decided to commission a thorough survey of the Cathedral, and Brunel was again invited to give his advice. But whatever the repairs involved, they do not appear to have required any further correspondence, and the main interest of the exchange is the light which it throws on Brunel's methods – and hours – of work.

Other Bristol commitments included some measure of involvement in the construction of the hotel behind the present Council House: it is supposed to have been built to accommodate transit passengers betweeen the GWR and the S.S. *Great Western*, and it has been converted into offices with the name 'Brunel House'. Brunel also encouraged several local industries with orders. Stephen Moulton's india rubber springs, then being manufactured at Bradford-on-Avon, and William Butler's creosote, being prepared as a timber preservative at his Crew's Hole factory, both received Brunel's attention. And in contemplating the future development of the Bristol region, Brunel perceived the importance of a rail crossing over the Severn estuary by a bridge or a tunnel. Writing of the possibility of a Severn Bridge at Old Passage he said:

I believe firmly that before 50 years are over there will be one (or a tunnel).[58]

The bridge took rather more than a hundred years to materialize, but the Severn Tunnel was operating in just over thirty years so on balance the forecast was reasonably accurate.[59]

Brunel showed similar perception of the long term needs of the region by his advocacy of deep water docking facilities in positions more accessible than the Floating Harbour. A scheme had been canvassed for a 'floating pier' in the early 1830s, and W. C. Mylne carried out an engineering survey for this (the beautifully finished drawings survive in the Port of Bristol Authority archives). The scheme was revived and modified at the end of the decade, when Brunel submitted a long report to a committee investigating possible port improvements. He warmly recommended the project for a pier at Portishead in a bid to re-capture the Bristol leadership in trans-oceanic trade, and concluded:

> . . . it is now for you to determine whether, by a comparatively easy, if prompt, effort you will place Bristol in that position, while there is yet time to command a decided preference in the establishment of such a trade, which can afterwards be easily maintained, and which must lead to an immense extension: but which, if now lost, and if the present opportunity be allowed to escape, may never be regained.[60]

The opportunity was not taken, but five years later Brunel put forward a similar scheme which led to the establishment of the Portbury Pier and Railway Company with a capital of £200,000. However, as this involved opening up the Somerset side of the Clifton Gorge by a railway line it caused some anxiety amongst the trustees of the Clifton Bridge. Brunel sought to reassure one of them that he was a loyal son of Bristol:

> I have heard from Osborne of the difficulty arisen as to what I may call my *scheme* – however altho *paternal* affection made me determined to do everything I could for Clifton Bridge which must and shall be finished *filial* affection for my parent shall not fail me – Bristol interests must be first thought of – and if one can't have a Portishead line that will benefit the bridge – we must have some other – for I quite agree with you that all hands must pull together for our revered parent the City of Bristol – and I am always ready to help.[61]

The resources of this venture appear soon to have been exhausted, as Brunel submitted an account in February 1849 for £674. 13. 1d., 'up to the time of the suspension of the works', and Latimer records that the project was formally abandoned in 1852.[62] Brunel was clearly exasperated by this collapse, writing to a correspondent who pressed him for information about the scheme:

Sir – I must beg you not to address me letters on the subject of your affairs with the Portishead Ry. Co. I never had anything to do with the matter. I know nothing about it and cannot be bothered with it. . . .[63]

So the matter slumbered for another decade. The railway to Portishead was eventually opened in 1867 and the pier in 1868, but the latter was soon made redundant by the construction of Portishead Dock. This dock, together with those at Avonmouth, at last gave Bristol the deep-water dock facilities which it needed in order to maintain a competitive port in the nineteenth century. Brunel's vision of Bristol's development was thus fulfilled, but only after his death.

Above Saltash Bridge under construction, 1858/9. Brunel's last work, like his first, was a bridge. He supervised the lifting of the first arch-span in 1857 but was so unwell with his final illness that the work was completed by R. P. Brereton.

Right A romantic view by night, from a 1930s picture postcard.

Robert Pearson Brereton, who worked with Brunel from
1836 and became his chief assistant in 1850.
Lithograph: Brunel Society.

Saltash Bridge today. This single track railway bridge
used iron tubes first tried in the Bristol Docks
Cumberland Basin footbridge. The chains were those
originally made for the Clifton Suspension Bridge. It was
a tragedy that the dying Brunel was too sick to witness the
opening by Prince Albert on May 3rd, 1859.

Photo John Cornwell.

Overall length	2,200 ft.
Two main spans	415 ft.
Elliptical tubes	16 ft. 9 in. wide, 12 ft. 3 in deep

Brunel at the end of his career. Worn out by work and worry, he leans on a cane before one of the funnels of the *Great Eastern*. This photograph was taken ten days before his death.

Photo Brunel University Library, Uxbridge.

Sculptor John Doubleday putting the finishing touches to a statue of Brunel before it is cast in bronze. The completed statue now stands outside the Broad Quay, Bristol head office of Bristol & West Building Society. Another version (this time, seated) was commissioned by the Society for Paddington Railway Station, and was unveiled on the same day, Wednesday, May 26th, 1982.

Photo Edward Davies.

Conclusion

It is probably not exaggerating to say that I. K. Brunel made a greater contribution to the landscape of the Bristol region than any other single individual before or since, with his splendid bridge over the Clifton Gorge; with the railway network radiating from Bristol, and the original Temple Meads station still intact at its centre; with the Floating Harbour surviving largely because of the improvements which he introduced; and with the S.S. *Great Britain* now being handsomely restored in the dry dock from which she was launched in 1843. Most of Brunel's work in Bristol had been done by 1848, although at that time the piers of Clifton Bridge stood gaunt and unfinished.

The twenty years 1828–48 were certainly the most creative in the relationship between Brunel and Bristol. By 1848, the group of intimate friends on whom the success of the relationship depended had begun to disperse. Roch had already gone into retirement in Pembroke. Guppy removed to Italy in 1849 after the collapse of the Great Western Steamship Company, and established an engineering business in Naples. Claxton remained heavily involved in Brunel's enterprises, but left Bristol. So were loosened the bonds of personal affection which had tied Brunel closely to Bristol and which made him proud to regard himself as 'a Bristol man', and other preoccupations removed him still further from the affairs of the city.

To the end, however, he retained his interest in Bristol and proclaimed his readiness to assist 'the spirited merchants of Bristol' in any worthwhile enterprise. The trajectory of his meteoric career took him away from Bristol, so that after 1848 he found that his London office became the focus of his work much more than in the earlier years of comparatively peripatetic railway building. His visits to Bristol were few and short in the 1850s. But for all the hectic activity of his professional career in these years, it seems that Brunel continued to think of himself as an adopted son of Bristol, retaining a special relationship of trust and affection with the city.

And by and large, Bristol responded by recognizing Brunel as one of its favourite sons. If Latimer's views accurately reflect the feelings of some late nineteenth century Bristolians about Brunel, counting the cost of the broad gauge when it was decided to convert it to standard gauge, it is little credit to their powers of discernment. With the advantage of greater hindsight it is possible to perceive the wisdom of Daniel Gooch's valedictory remarks on Brunel:

By his death the greatest of England's engineers was lost, the man with the greatest originality of thought and power of execution, bold in his plans but right. The commercial world thought him extravagant; but although he was so, great things are not done by those who sit down and count the cost of every thought and act.[64]

Posterity has decided in favour of the judgment of Gooch rather than that of Latimer, and however much his projects might have cost their ancestors, Bristolians are generally happy now to acknowledge I. K. Brunel as an adopted citizen, welcoming the erection of his statue in the heart of the city. He is certainly one of the handful of masterminds whose imagination helped to mould the city and one of those in whose works the city can still take pride.

Notes and References

1. J. Latimer: *Annals of Bristol in the Nineteenth Century*, Bristol 1887, p.191.

2. The diaries and other papers of Sir Marc Brunel are in the Library of the Institution of Civil Engineers, and I am grateful to the Librarian of the Institution, Miss D. J. Bayley, for permission to consult them.

3. The standard biographies are I. Brunel: *The Life of Isambard Kingdom Brunel*, 1870; Celia Brunel Noble: *The Brunels, Father and Son*, 1938; and L. T. C. Rolt: *Isambard Kingdom Brunel*, 1957. The biography of 1870 does not mention a visit to Clifton before the bridge competition of 1829, but both Lady Noble and Mr. Rolt refer to a period of recuperation in Clifton, presumably in 1828. Both the latter refer to manuscript diaries of I. K. Brunel which I have not had the good fortune to see.

4. PLB, Brunel to J. N. Miles, April 28th 1845.

5. PLB, Brunel to Osborne, June 3rd 1845.

6. PLB, Brunel to John Yates, November 16th 1854.

7. PLB, Brunel to N. Roch, May 10th 1844.

8. PLB, Brunel to Brodie, October 28th 1844.

9. PLB, Brunel to N. Roch, November 8th 1844.

10. Susan Thomas: *The Bristol Riots*, Bristol Branch of the Historical Association, 1974, p.1.

11. *Ibid.*, p.26.

12. See Latimer, *op. cit.* pp.206–18, for an account of municipal reform in Bristol.

13. See Rolt, *op. cit.* pp.59–63, for a reconstruction of Brunel's possible movements during the Bristol Riots.

14. Brunel to Hawes, March 27th 1831, quoted Noble, *op. cit.* p.109 and Rolt, *op. cit.* p.56.

15. Quoted Rolt, *op. cit.* p.58.

16. Latimer, *op. cit.* p.133.

17. PLB, Brunel to Ward, May 10th 1849.

18. PLB, Brunel to Ward, October 19th 1849.

19. PLB, Brunel to R. Palmer, July 16th 1851.

20. PLB, Brunel to G. Hennet, December 6th 1851.

21. There is an extensive literature on the Clifton Bridge. For a recent discussion of its technical features, see R. F. D. Porter Goff: 'Brunel and the design of the Clifton Suspension Bridge', *Proceedings Institution of Civil Engineers* Pt. I, August 1974. As a more popular account, see G. Body: *Clifton Suspension Bridge*, 1975.

22. The report is in the Port of Bristol Authority (PBA) archives. For a fuller account, see R. A. Buchanan: 'I. K. Brunel and the Port of Bristol', *Transactions of the Newcomen Society*, vol. 42, 1969–70, pp.41–56.

23. PBA archives: letter from Brunel, June 3rd 1844.

24. PLB, Brunel to C. Claxton, June 3rd 1844.

25. R. A. Buchanan, *op. cit.*, and also 'Cumberland Basin, Bristol', *Industrial Archaeology*, vol. 6 no. 4, November 1969.

26. Bristol Docks Company (BDC) Minutes, April 27th 1846.

27. PLB, Brunel to Osborne, May 1st 1846.

28. Docks Committee, Bristol Corporation. Minutes September 4th 1848.

29. PLB, Brunel to Bell, October 6th 1848 and December 27th 1849. Bell was Brunel's Assistant who became responsible for completing the work after the death of Hammond.

30. PLB, Brunel to Osborne, October 25th 1848.

31. PLB, Brunel to Rennie, January 16th 1852, and Brunel to Richard Poole King, October 21st 1852.

32. PLB, Brunel to J. Gibbs, November 8th 1852 and November 20th 1852.

33. PLB, Bennett (Clerk to Brunel) to L. Bruton, August 17th 1858.

34. Brunel's Report to the BDC, June 1844, in PBA archives.

35. Latimer *op. cit.* p.247.

36. Quoted, Celia Noble, *op. cit.* p.116.

37. The standard account is E. T. MacDermot: *History of the Great Western Railway*, 1927, revised ed. by C. R. Clinker, 1964.

38. *Bristol Times and Bath Advocate* August 18th 1849.

39. *Bristol Times and Mirror* January 28th 1871.

40. PLB, Brunel to J. Gibbs, August 11th 1846.

41. PLB, Brunel to G. Jones, April 18th 1844.

42. PLB, Brunel to A. Jamieson, May 6th and 9th 1844.

43. PLB, Brunel to C. Richardson, September 14th 1858.

44. PLB, Brunel to C. Richardson, September 14th 1853.

45. PLB, Brunel to C. Richardson, September 17th 1858.

46. See R. A. Buchanan and N. Cossons: *Industrial Archaeology of the Bristol Region*, Newton Abbot 1969, for a general account of these railway works. For a more specialized treatment, see J. W. Totterdill, 'A peculiar form of construction', *Journal of the Bristol & Somerset Society of Architects*, vol. 5, pp.111–12; and J. Mosse, 'Bristol Temple Meads', *BIAS Journal* 4, 1971.

47. See Grahame Farr: *The Steamship Great Western*, Bristol Branch of the Historical Association, 1963.

48. PLB, Brunel to Hunt, December 11th 1844.

49. See Grahame Farr: *The Steamship Great Britain*, Bristol Branch of the Historical Association, 1965; and Ewan Corlett: *The Iron Ship*, Moonraker Press, Bradford-on-Avon, 1975. Also Buchanan and Cossons, *op. cit.* p.49.

50. PLB, Brunel to Society of Merchant Venturers, September 20th and October 4th 1842.

51. Latimer, *Annals* p.281, and *History of the Society of Merchant Venturers*, p.264, although he does not attribute it in either work to Brunel.

52. PLB, Brunel to C. Savery, March 23rd 1846.

53. Latimer, *Annals*, p.281n.

54. PLB, Brunel to Jas. Simpson, November 30th 1848.

55. See R. A. Buchanan: *Nineteenth century engineers in the Port of Bristol*, Bristol Branch of the Historical Association, 1971.

56. PLB, Brunel to the Revd. E. Banks, June 1st 1850. The Dean at this time was G. E. May, who had just begun a term of office which was to last until 1891.

57. PLB, Brunel to the Revd. Dean of Bristol, June 24th 1850.

58. PLB, Brunel to J. Hooper, May 30th 1854.

59. The Severn Tunnel was open to traffic in 1886, and the Severn Bridge in 1966. But of course the Sharpness Bridge, further up the Severn, was open in 1879: this was put out of action when a barge struck one of the piers in 1960 and was dismantled in 1967.

60. Report by Brunel, printed by order of the Committee, December 26th 1839. Latimer, *Annals* p.397, curiously attributed the scheme to John MacNeil.

61. PLB, Brunel to J. N. Miles, April 28th 1845.

62. PLB, Bennett to F. R. Ward, February 22nd 1849, and Latimer, *Annals*, p.397.

63. PLB, Brunel to W. Clarke, July 6th 1853.

64. Sir Daniel Gooch: *Memoirs & Diary*, Newton Abbot, 1972, p.76.

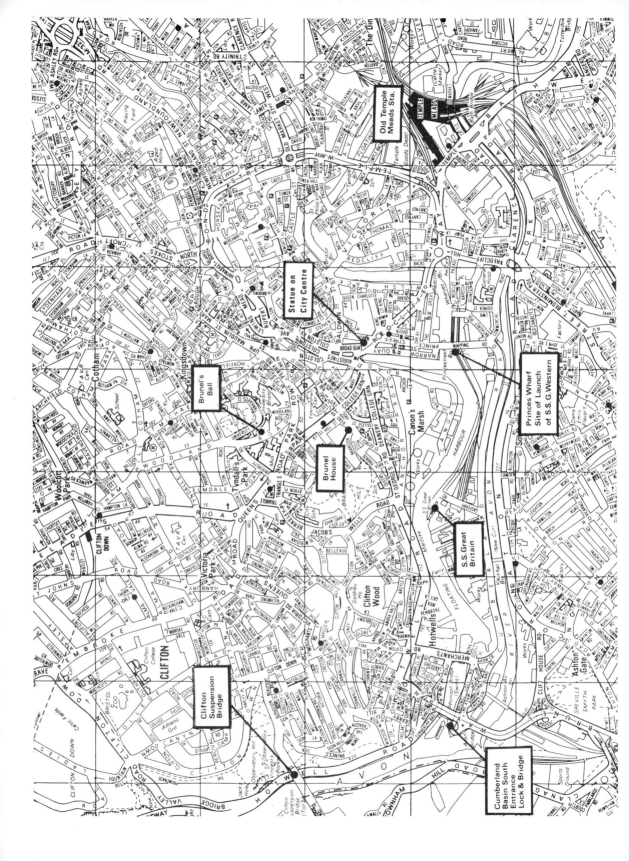

Old Temple Meads Sta.

Statue on City Centre

Brunel's Ball

Brunel House

Princes Wharf Site of Launch of S.S.G.Western

S.S. Great Britain

Clifton Suspension Bridge

Cumberland Basin South Entrance Lock & Bridge

Brunel sites in the City of Bristol

Residents of Bristol and its surrounding districts have the unique opportunity of visiting and re-visiting the Brunel sites, and witnessing the fascinating restoration in progress on the S.S. *Great Britain*. An enthusiastic and energetic visitor can manage to see all in one day, but the short stay visitor is recommended to restrict himself to the major attractions. The Suspension Bridge demands an hour for a leisurely walk to and fro across it and a stroll to admire the dramatic views from Observatory Hill (see map). The *Great Britain* and its Museum ideally require at least two hours.

CLIFTON SUSPENSION BRIDGE is on the southern end of Clifton Down. It can be reached by car along the Portway, up Bridge Valley Road turning right at the top of the hill following the signpost. An alternative route is up Whiteladies Road/Blackboy Hill turning left on reaching the Downs. Car Parking space can be found in the adjacent streets. Buses from Temple Meads or the Centre — Colston Avenue East — go up to Christchurch Green.

BRUNEL HOUSE in St. George's Road was once Brunel's Royal Western Hotel. Only the façade is original. The building is behind the Council House and College Green and but a short walk from the City Centre.

THE BRUNEL STATUE a larger than life-size bronze by John Doubleday, stands outside the Bristol & West Building Society Head Office extension on Broad Quay on the City Centre.

TEMPLE MEADS TERMINUS AND ENGINE SHED is on Temple Gate, one of the busiest and noisiest thoroughfares in Bristol. There may be limited car parking in the station precinct but there is a convenient N.C.P. indoor car-park opposite from where the station can be reached by footbridge. Buses run from Clifton and the Centre for the Station.

WAPPING WHARF was alongside Wapping Road on the south side of Prince Street Bridge. A plaque commemorates the site of the launch of the *Great Western*. The Bristol Industrial Museum is nearby. This site is passed on the journey from the City Centre to the *Great Britain*.

THE S.S. GREAT BRITAIN is in Gas Ferry Road signposted off Cumberland Road. There is plenty of free car-parking space available. Not conveniently reached by bus, the ship is within walking distance of the Centre along the water front on the Industrial Museum side of the harbour. The ship is open all year round except Christmas Eve and Christmas Day. Opening hours are 10.00 a.m. to 5.00 p.m. in the winter and 10.00 a.m. to 6.00 p.m. in the summer. A museum and souvenir shop are on the site and refreshments are available.

CUMBERLAND BASIN WITH BRUNEL'S SOUTH ENTRANCE LOCK AND TUBULAR IRON FOOTBRIDGE. Perhaps of appeal to enthusiasts only. There is plenty of car-parking space available and good views of Clifton and the Suspension Bridge.

BRUNEL'S BALL is a massive boulder which stands on Woodland Road and is a short walk from the University Tower. It was one of two natural nodules that were dug out in 1837 during the construction of a Great Western Railway tunnel at St Anne's. Brunel had the pair mounted on pedestals at one end of the tunnel.

Useful Addresses

Bristol City Museum and Art Gallery, Queens Road, Bristol BS8 1RL. *Tel. (0272) 223571.* Opening hours 10 a.m. - 5 p.m. Daily. The Museum has a collection of Brunel material (prints, photographs, drawings, contemporary lithographs and aquatints etc.). Research enquiries should be made by telephone or letter to the Curator of Technology, Bristol Industrial Museum, Princes Wharf, Prince Street, Bristol.

Bristol Industrial Museum, Princes Wharf, Prince Street, Bristol. *Tel. (0272) 251470.* The Museum is open from Saturday to Wednesday (open on Sunday) from 10 a.m.–5 p.m. (closed 1-2) and includes a Brunel exhibition.

S.S. Great Britain Project, Great Western Dock, Gas Ferry Road, bristol BS1 6TY. *Tel. (0272) 260680.* Open Monday to Sunday, 10 a.m.–6 p.m. summertime, 10 a.m.–5 p.m. wintertime. Closed Christmas Eve and Christmas Day only. Apart from the historic dock which you can inspect you are able to go aboard the ship in its present stage of restoration. There is a dockside museum, a souvenir shop and a snack-bar.

University of Bristol Library, Tyndall Avenue, Bristol BS8 1TJ. *Tel. (0272) 303030.* The University houses a special collection of Brunelia, including letter books, sketch books, memoranda, reports, calculations and surveys covering the period from 1830 to 1859. There is a display of Brunel's drawing instruments, small tools, rules, set squares, a barometer and binoculars, which were owned by Brunel and his sons. These may be inspected by appointment with the Special Collections Librarian.

Recommended Reading

BALL Adrian & WRIGHT Diana   *S.S. GREAT BRITAIN,* David & Charles 1981

BECKETT Derrick   *BRUNEL'S BRITAIN,* David & Charles 1979

BLAKE Joe   *RESTORING THE GREAT BRITAIN,* Redcliffe Press Ltd 1989

BODY Geoffrey   *CLIFTON SUSPENSION BRIDGE,* Moonraker Press 1976

BRUNEL Isambard   *THE LIFE OF ISAMBARD KINGDOM BRUNEL,* David & Charles Reprint 1971

BUCHANAN R.A. & COSSONS Neil   *THE INDUSTRIAL ARCHAEOLOGY OF THE BRISTOL REGION,* David & Charles 1969

CORLETT Ewan   *THE IRON SHIP,* Moonraker Press 1975

PUDNEY John   *BRUNEL AND HIS WORLD,* Thames and Hudson 1974

PUGSLEY Sir Alfred   *THE WORKS OF ISAMBARD KINGDOM BRUNEL,* Institution of Civil Engineers, London and University of Bristol 1976

ROLT L. T. C.   *ISAMBARD KINGDOM BRUNEL,* Longmans Green 1957